PERENNIAL MATH

CONTEST PROBLEMS
2014-2017

Dr. Sylvia Dean was a gifted specialist and administrator for 26 years in Huntsville, Alabama. She created Perennial Math in 2011 upon completion of her doctoral degree. She has hosted many robotics, technology, and mathematics competitions throughout the years and has tirelessly worked on providing educational opportunities to students around the world. It has always been her ambition to leave students with the feeling that "math is cool." These problems are challenging and students should know up front that scoring a 3 out of 6 on any of these tests is extraordinary and above average. Great problem-solvers learn persistence and appreciation for challenge. Good luck as you encourage your students to take the quest to become great mathletes.

INTERMEDIATE LEVEL

Grades 5-6

Name________________________________

PERENNIAL MATH

TEST # 1

November 2014 Intermediate Level

1) In this multiplication problem, the stars stand for the same digit. What is the largest possible three-digit product?

2) Mary is 11 years old. Next year she will be three times as old as her brother will be. How many years from now will she be twice as old as her brother will be then?

3) Billy is playing in a single elimination tennis tournament which means once a player loses he or she must go home. If there are 32 players, how many games will he have to win in order to be the champion?

4) Four girls ran for homecoming queen. Lori got 60% of the votes, Mattie got 25%, Kendra got 10% and Sonya got the remaining 45 votes. How many total votes were cast altogether?

5) If you start at the bottom left point and go to the top right point, how many paths along the lines can you draw? You may only travel up or right at any time. You may not move down or left.

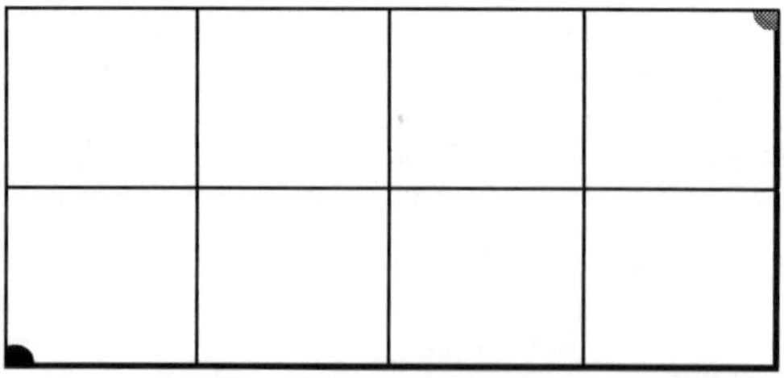

6) Farmer Joe counted the horses and chickens in his barn. He counted 19 heads and 50 feet. How many chickens were in the barn?

PERENNIAL MATH

TEST #1

November 2014 Intermediate Level Answers

1) In this multiplication problem, the stars stand for the same digit. What is the largest possible three-digit product? 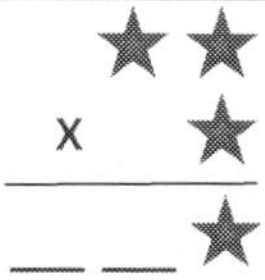	396
2) Mary is 11 years old. Next year she will be three times as old as her brother will be. How many years from now will she be twice as old as her brother will be then?	5
3) Billy is playing in a single elimination tennis tournament which means once a player loses he or she must go home. If there are 32 players, how many games will he have to win in order to be the champion?	5
4) Four girls ran for homecoming queen. Lori got 60% of the votes, Mattie got 25%, Kendra got 10% and Sonya got the remaining 45 votes. How many total votes were cast altogether?	900
5) If you start at the bottom left point and go to the top right point, how many paths along the lines can you draw? You may only travel up or right at any time. You may not move down or left. 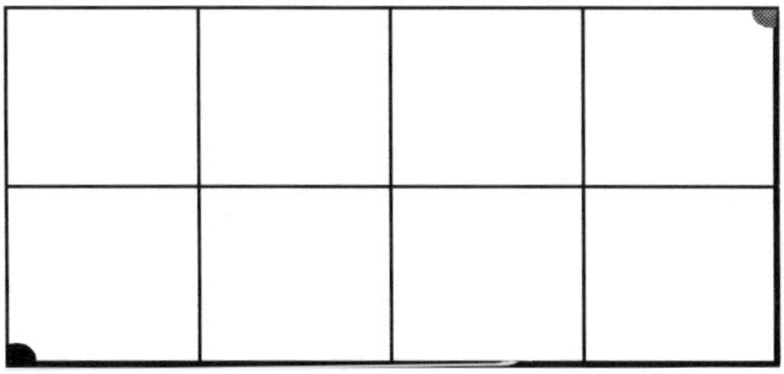	15
6) Farmer Joe counted the horses and chickens in his barn. He counted 19 heads and 50 feet. How many chickens were in the barn?	13

PERENNIAL MATH

TEST # 1

November 2014 Intermediate Level Solutions

We are only providing some suggested solutions below. There are always multiple methods to solve problems.

1) Since the star stands for the same digit, the star times star equals star. The only digits that are capable of this are 0, 1, 5 and 6. Six is the largest, so 66 x 6 = 396. That is the largest product you can make.

396

2) If Mary is 11 now, then next year she will be 12. Next year she will be three times as old as her brother, so he will be 4 next year. Make a chart to see the ages until you notice when she is twice as old as her brother. In 5 years she will be 16 and he will be 8.

	Now	in 1 year	in 2 years	in 3 years	in 4 years	in 5 years
Mary's age	11	12	13	14	15	16
Brother's age	3	4	5	6	7	8

5

3) Make a tree diagram so you can see the tournament bracket. Starting with 32 players we go to 16 player, 8 player, 4 players, 2 players and then the winner is left. The winner must play 5 games overall.

5

PERENNIAL MATH TEST #1

November 2014 Intermediate Level Solutions

4) Method 1: If you add up the percentages of three girls you have 60 + 25 + 10 = 95%. This leaves Sonya receiving the remaining 5% of votes which is 45 votes. If 45 votes is 5%, then 10% is 90 votes. So 100% is ten times this amount or 900 votes.

Method 2: Use algebra. Since Sonya has 5% of the total, let X represent the total.

$$.05X = 45$$
$$X = 45 / .05$$
$$X = 900$$

900

5)

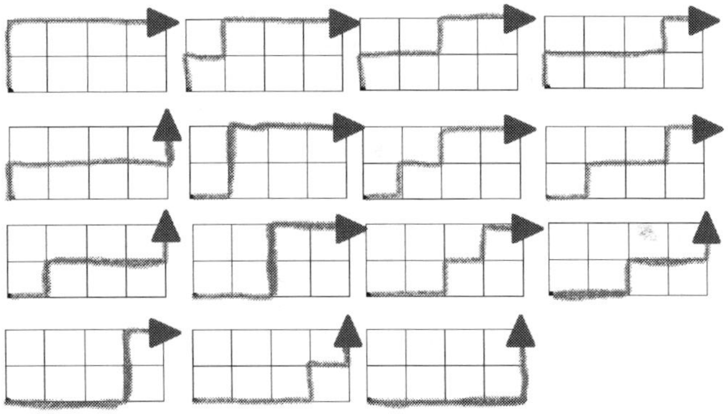

15

6) Make a table and see the relationship of animals and feet. Notice that with each decrease in horses, you lose 2 feet. There are 13 chickens in the barn.

Decreases by 1

Chickens	10	11	12	13
Horses	9	8	7	6
Total heads	19	19	19	19
Total feet	56	54	52	**50**

Decreases by 2

13

Name________________________________

PERENNIAL MATH

TEST #2

December 2014 Intermediate Level

1) In the following number series two terms are missing. What is the sum of the two numbers that should be in the blank spaces?

2, 3, 5, 7, 11, __, __, 19

2) Randy has pennies, nickels, and dimes. How many ways can he combine his coins to make 56 cents?

3) Mary gave Mike half of her gumdrops. Mike ate 1/3 of these gumdrops and gave the rest to Mel. Mel kept 6 of the gumdrops and gave the last 18 to Mark. How many gumdrops did Mary have to start with?

4) The eight players on the soccer team decided to race. The races only consisted of two members running against each other at any given time with the others watching. Each member raced each of the other members only once. How many races took place?

5) Using only the digits 1 through 9, place each digit along the three sides of the triangle in the circles so that each side adds up to 20. What is the sum of the three numbers on the corners?

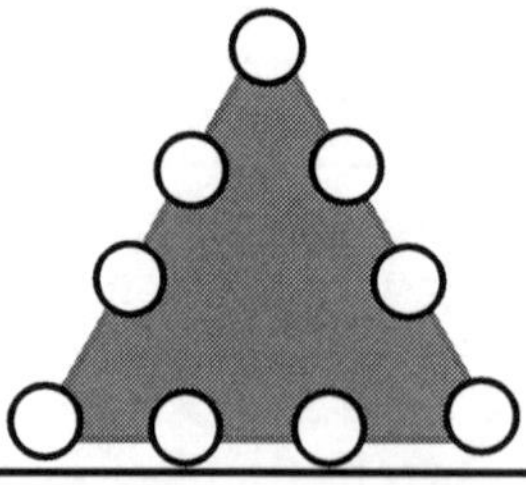

6) Bridgette had a square piece of paper that had an area of 36 cm^2. She cut out the largest rectangle she could that had a length three times longer than its width. What was the perimeter of this rectangle?

PERENNIAL MATH TEST #2

December 2014 Intermediate Level Answers

1) In the following number series two terms are missing. What is the sum of the two numbers that should be in the blank spaces? 2, 3, 5, 7, 11, ___, ___, 19	**30**
2) Randy has pennies, nickels, and dimes. How many ways can he combine his coins to make 56 cents?	**42**
3) Mary gave Mike half of her candy canes. Mike ate 1/3 of these candy canes and gave the rest to Mel. Mel kept 6 of the candy canes and gave the last 18 to Mark. How many candy canes did Mary have to start with?	**72**
4) Santa's eight reindeer decided to race. The races only consisted of two reindeer running against each other at any given time with the others watching. Each reindeer raced each of the other reindeer only once. How many races took place?	**28**
5) Using only the digits 1 through 9, place each digit along the three sides of the triangle in the circles so that each side adds up to 20. What is the sum of the three numbers on the corners? 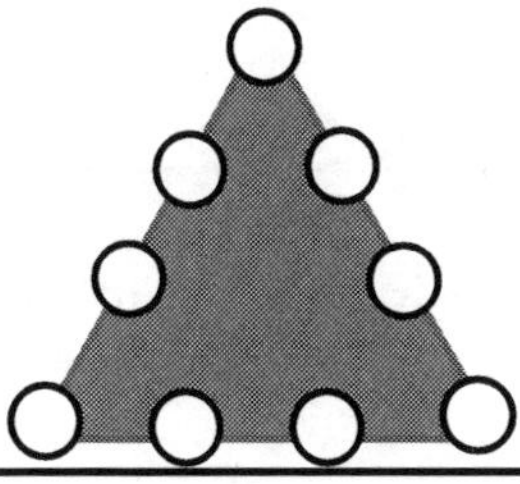	**15**
6) Bridgette had a square piece of wrapping paper that had an area of 36 cm^2. She cut out the largest rectangle she could that had a length three times longer than its width. What was the perimeter of this rectangle?	**16**

PERENNIAL MATH

TEST # 2

December 2014 Intermediate Level Solutions

We are only providing some suggested solutions below. There are always multiple methods to solve problems.

1) 2, 3, 5, 7, 11, __, __, 19 The sequence is a list of prime numbers and the two missing numbers are 13 and 17. Their sum is 30.

30

2) Make an organized list starting with the most amount of the highest coins and working down. There are 42 ways to make 56¢.

Dimes	Nickels	Pennies
5	1	1
5	0	6
4	3	1
4	2	6
4	1	11
4	0	16
3	5	1
3	4	6
3	3	11
3	2	16
3	1	21
3	0	26
2	7	1
2	6	6

Dimes	Nickels	Pennies
2	5	11
2	4	16
2	3	21
2	2	26
2	1	31
2	0	36
1	9	1
1	8	6
1	7	11
1	6	16
1	5	21
1	4	26
1	3	31
1	2	36

Dimes	Nickels	Pennies
1	1	41
1	0	46
0	11	1
0	10	6
0	9	11
0	8	16
0	7	21
0	6	26
0	5	31
0	4	36
0	3	41
0	2	46
0	1	51
0	0	56

42

3) Work backwards. Mark has the last 18 and we can add 6 that Mel kept which is 24 candy canes. 24 was 2/3 of what Mike had, so Mike had 36 and that was 1/2 of what Mary had to start with. So Mary had 72 candy canes to start with. Work it forward to check:

$(72 \div 2 \longrightarrow 36)$ $(36 \div 3 \longrightarrow 12)$ $(24 - 6 \longrightarrow 18)$

Mary *Mike* *Mike eats 12* *Mel* *Mark*

72

PERENNIAL MATH

TEST #2

December 2014 Intermediate Level Solutions

4) Make a chart to see the pairings.
In this combination, there are 7 + 6 + 5 + 4 + 3 + 2 + 1 = 28 races.

Reindeer 1	Reindeer 2	Reindeer 3	Reindeer 4	Reindeer 5	Reindeer 6	Reindeer 7
Reindeer 2	Reindeer 3	Reindeer 4	Reindeer 5	Reindeer 6	Reindeer 7	Reindeer 8
Reindeer 3	Reindeer 4	Reindeer 5	Reindeer 6	Reindeer 7	Reindeer 8	
Reindeer 4	Reindeer 5	Reindeer 6	Reindeer 7	Reindeer 8		
Reindeer 5	Reindeer 6	Reindeer 7	Reindeer 8			
Reindeer 6	Reindeer 7	Reindeer 8				
Reindeer 7	Reindeer 8					
Reindeer 8						

28

5) There are 6 different arrangements that will yield a sum of 20 on each side. Regardless to which arrangement you use, the corner sums will always yield 15.

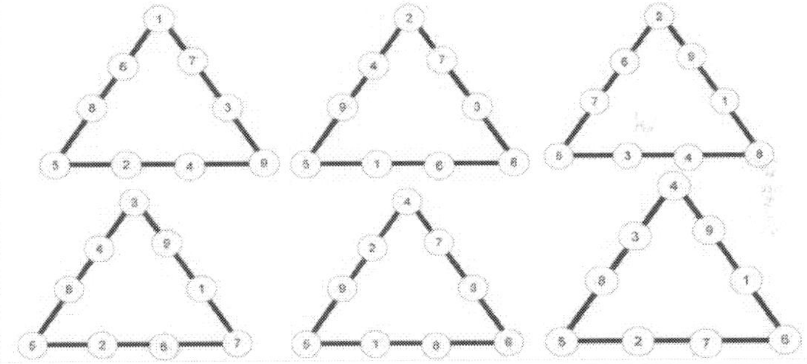

15

6) Area equals length times width. So the dimensions of the square with $36m^2$ are: 6m x 6m. The largest rectangle that could have been cut from this with a length three times the width is 6m x 2m.
Perimeter is equal to 2L + 2W.

2(6m) + 2(2m) = 12 + 4 = 16m.

Therefore, the perimeter of this rectangle is 16cm.

16

Name______________________________

PERENNIAL MATH

TEST #3

January 2015 Intermediate Level

1) What is the least natural number that has three different prime factors?	
2) Jenny was ordering cans of waffle mix and they costs \$5.19 per can or a 3-pack costs \$13.57. She ordered 16 cans and paid \$9.99 shipping. She calculated to find the least expensive way to order and paid with a \$100 gift card. How many dollars are left on the gift card if you round the amount to the nearest dollar?	
3) A palindrome is a number that reads the same forwards and backward such as 121. How many palindromes are there between 125 and 521?	
4) There are 3 drawers in the office file cabinet. The number of files in the second drawer equals the sum of the number of files in the top and bottom drawers combined. Donna takes 8 files from the top drawer and puts them on the desk. She also takes 4 files from the bottom drawer and puts them in the second drawer and then takes 6 files from the second drawer and puts them in the top drawer. How many more files are now in the second drawer than in the top and bottom drawers combined?	
5) In this cryptarithm, different letters stand for different digits. There are several solutions for this problem. What is the largest number that ACD can be? $\begin{array}{r} BA \\ +\ BA \\ \hline ACD \end{array}$	
6) In a class of 28 students, 15 like chocolate ice cream and 18 like vanilla ice cream. However, 3 students do not like these flavors. How many students in the class like both chocolate and vanilla?	

PERENNIAL MATH

TEST # 3

January 2015 Intermediate Level Answers

1) What is the least natural number that has three different prime factors?	**30**
2) Jenny was ordering cans of waffle mix and they costs \$5.19 per can or a 3-pack costs \$13.57. She ordered 16 cans and paid \$9.99 shipping. She calculated to find the least expensive way to order and paid with a \$100 gift card. How many dollars are left on the gift card if you round the amount to the nearest dollar?	**17**
3) A palindrome is a number that reads the same forwards and backward such as 121. How many palindromes are there between 125 and 521?	**39**
4) There are 3 drawers in the office file cabinet. The number of files in the second drawer equals the sum of the number of files in the top and bottom drawers combined. Donna takes 8 files from the top drawer and puts them on the desk. She also takes 4 files from the bottom drawer and puts them in the second drawer and then takes 6 files from the second drawer and puts them in the top drawer. How many more files are now in the second drawer than in the top and bottom drawers combined?	**4**
5) In this cryptarithm, different letters stand for different digits. There are several solutions for this problem. What is the largest number that ACD can be? $\begin{array}{r} B\,A \\ +\ B\,A \\ \hline A\,C\,D \end{array}$	**182**
6) In a class of 28 students, 15 like chocolate ice cream and 18 like vanilla ice cream. However, 3 students do not like these flavors. How many students in the class like both chocolate and vanilla?	**8**

PERENNIAL MATH TEST #3

January 2015 Intermediate Level Solutions

We are only providing some suggested solutions below. There are always multiple methods to solve problems.

1) The least natural number that has three different prime factors can be solved by listing the first three prime numbers and multiplying: 2 x 3 x 5 = 30

30

2) The least amount Jenny can spend is 5 times the 3-pack plus one additional can for the $5.19. So 5 x $13.57 = $67.85 + $5.19 = $73.04. Add the shipping $73.04 + 9.99 = $83.03. If the bill is paid with a $100 gift card, then $100 - $83.03 leaves $16.97 on the gift card. This number rounded up is $17

17

3) Make an organized list of palindromes between 125 and 521. There are 7 that begin with 1, and 10 each that begin with 2, 3, and 4. And finally there are 2 that begin with 5. So, there are 39 palindromes between 125 and 521.

131	202	303	404	505
141	212	313	414	515
151	222	323	424	***2***
161	232	333	434	
171	242	343	444	
181	252	353	454	
191	262	363	464	
7	272	373	474	
	282	383	484	
	292	393	494	
	10	***10***	***10***	

39

PERENNIAL MATH

TEST #3

January 2015 Intermediate Level Solutions

4) When 8 files are removed from the top drawer, then there are 8 more in the second drawer than the top and bottom combined. When 4 are taken from the bottom drawer and put in the second drawer (a net change of 8) there are now 16 more in the second drawer than the top and bottom. When 6 are taken from the second drawer and placed in the top drawer (a net change of 12), there are now 4 more in the second cabinet than in the top and bottom.

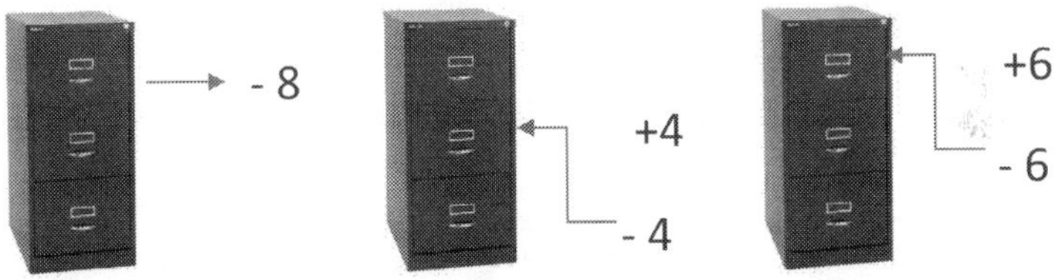

4

5) There are several possibilities for this cryptarithm. However, A must be 1 or 2 since it is the result of a carry on the addition of the B+B. Some of the possible solutions are 102, 142, 162, and 182. The answer is 182 since it is the largest possible sum.

$$\begin{array}{r} B\,A \\ +\;B\,A \\ \hline A\,C\,D \end{array}$$

182

6) Use algebra. Let n represent the region for both flavors.

$$28 = (15 - n) + (18 - n) + 3 + n$$
$$28 = 36 - n$$
$$n = 8$$

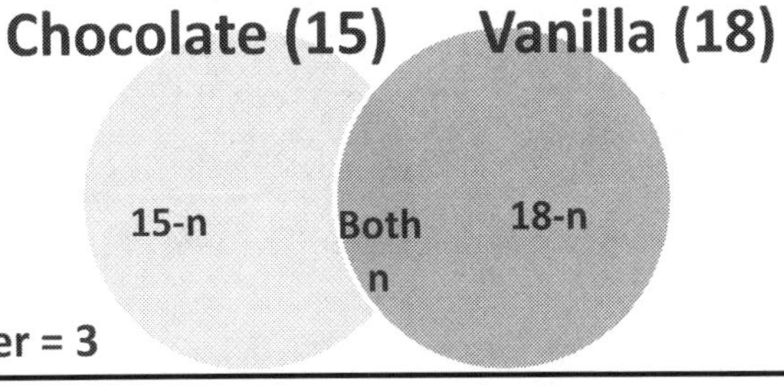

Neither = 3

8

Name________________________________

PERENNIAL MATH

TEST #4

February 2015 Intermediate Level

1) One natural number is 3 times larger than another natural number. The product of the two numbers is 27. What is the sum of the two numbers?

2) Vince is thinking of a four-digit number. The first average of the first two digits is 5. The average of the last two digits is 3. The average of the first and last digit is 4. What is the greatest number he could have been thinking of?

3) How many numbers between 100 and 300 are divisible by both 3 and 5?

4) Parents donated chocolate for the school fund raiser. Forty pounds of milk chocolate sold for $2.15 per pound and dark chocolate sold for $1.90 per pound. The class made $158.20. How many total pounds of chocolate were sold?

5) The counting numbers are written across the board in order. There is room for 50 digits across and there are no spaces between each digit. John writes two rows of digits. What is the last digit in the second row?

6) How many rectangles can be made whose sides are even whole numbers and they have a perimeter of 64 cm?

PERENNIAL MATH

TEST #4

February 2015 Intermediate Level Answers

1) One natural number is 3 times larger than another natural number. The product of the two numbers is 27. What is the sum of the two numbers?	12
2) Vince is thinking of a four-digit number. The first average of the first two digits is 5. The average of the last two digits is 3. The average of the first and last digit is 4. What is the greatest number he could have been thinking of?	8260
3) How many numbers between 100 and 300 are divisible by both 3 and 5?	13
4) Parents donated chocolate for the school fund raiser. Forty pounds of milk chocolate sold for $2.15 per pound and dark chocolate sold for $1.90 per pound. The class made $158.20. How many total pounds of chocolate were sold?	78
5) The counting numbers are written across the board in order. There is room for 50 digits across and there are no spaces between each digit. John writes two rows of digits. What is the last digit in the second row? 	5
6) How many rectangles can be made whose sides are even whole numbers and they have a perimeter of 64 cm?	8

PERENNIAL MATH

TEST # 4

February 2015 Intermediate Level Solutions

We are only providing some suggested solutions below. There are always multiple methods to solve problems.

1) The numbers 3 and 9 produce 27 when multiplied and 9 is also 3 times bigger than 3. The sum of the two numbers is 9 + 3 = 12

12

2) Since the average of the first two digits is five, the sum of the digits must add up to 10 - that leaves numbers of (1, 9), (2, 8), (3, 7), or (4, 6).
The average of the last two digits is 3, so the sum of the digits must add up to 6 - that leaves possible numbers of (1, 5), or (2, 4).
If the average of the first and last digit is four, then the sum of the digits must add up to 8 - that leaves possible numbers of (0,8), (1, 7), (2, 6), or (3, 5).
The greatest number possible can be found by choosing the highest numbers in the possible lists (0,8).
8_ _ 0 since 8 is the highest number in the averages of first and last numbers. If 8 is the first number then, 2 is the next number because the average of the first two numbers is 5.
82_0 Since 0 is the last number the third number must be 6 because the average of the last two digits is 3.
The number is 8260.

8260

3) The first number between 100 and 300 that is divisible by both 3 and 5 is 105. The least common multiple of 3 and 5 is 15. So add every 15th number past 105. The numbers are: 105, 120, 135, 150, 165, 180, 195, 210, 225, 240, 255, 270, 285. Do not count 300 because it says between 300. There are 13 numbers that are between 100 and 300 are divisible by both 3 and 5.

13

PERENNIAL MATH

TEST # 4

February 2015 Intermediate Level Solutions

4) Forty pounds of milk chocolate sold for $2.15 per pound so that was a total of $86. The class made a total of $158.20 altogether. So subtract the milk chocolate total to find out how much money was made from the dark chocolate. $158.20 - $86 = $72.20. Divide to find out how many pounds of dark chocolate were sold. $72.20/$1.90 = 38. 40 lbs. of milk chocolate + 38 lbs. of dark chocolate = 78 total lbs. of chocolate sold.

78

5) You could list the numbers out:
1234567891011121314151617181920212223242526272829303132333435363738394041424344454647484950515253545.

123456789101112131415161718192021222324252627282930
3132333435363738394041424344454647484950515253545.

The last number on the second row is 5.

Or you could calculate that the first 9 counting numbers is 9 digits, then all the rest of the counting numbers are 2-digit. So 100 digits in both rows minus the first 9 numbers is 91 digits. Divide 91 in half and that is 45.5
The numbers 10-19, 20-29, 30-39, 40-49 are 20 digits each. So the counting number 1-49 equals 89 digits. Eleven digits further would land you on 5, the beginning digit of the number 55.

5

6) Make a chart of the rectangles that can be made with dimensions that yield a perimeter of 64 cm, count only the ones with even numbers. There are 8 rectangles that have even dimensions.

Length	Width
31	1
30	2
29	3
28	4
27	5
26	6
25	7
24	8
23	9
22	10
21	11
20	12
19	13
18	14
17	15
16	16

8

Name________________________________

PERENNIAL MATH

TEST #1

November 2015 Intermediate Level

1) What single number should be placed in both boxes?

$$30 - \square = 22 + \square$$

2) Rachel has twice as much money as Kim. Kim has two dimes and two nickels. Rachel only has quarters and nickels and has the same number of each coin. If they trade all their money in for only nickels, how many nickels would they have together?

3) Milo spent half of his weekly allowance for a movie ticket. His parents let him wash the car for $15 to earn more money. How much is his weekly allowance if he ended with $24?

4) A palindrome is a number that reads the same forwards or backwards. 2002 is a palindrome. What is the sum of the closest palindrome before 2002 and the closest palindrome after 2002?

5) Each hexagon is made by adding the 2 numbers in the hexagons below it. Complete the puzzle. Take the number of the top hexagon and add 20 to it, what value do you get?

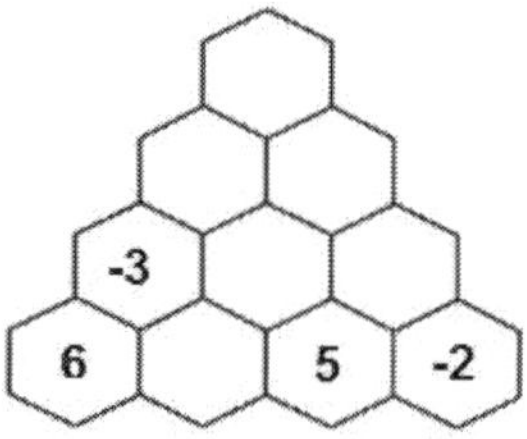

6) For each of the numbers: 41, 53, and 82, the tens digit is greater in value than the ones digit. How many 2-digit numbers have this property, but exclude a 6 in either digit?

PERENNIAL MATH TEST # 1

November 2015 Intermediate Level Answers

<table>
<tr><td>1) What single number should be placed in both boxes?

$$30 - \square = 22 + \square$$</td><td>4</td></tr>
<tr><td>2) Rachel has twice as much money as Kim. Kim has two dimes and two nickels. Rachel only has quarters and nickels and has the same number of each coin. If they trade all their money in for only nickels, how many nickels would they have together?</td><td>18</td></tr>
<tr><td>3) Milo spent half of his weekly allowance for a movie ticket. His parents let him wash the car for $15 to earn more money. How much is his weekly allowance if he ended with $24?</td><td>18</td></tr>
<tr><td>4) A palindrome is a number that reads the same forwards or backwards. 2002 is a palindrome. What is the sum of the closest palindrome before 2002 and the closest palindrome after 2002?</td><td>4103</td></tr>
<tr><td>5) Each hexagon is made by adding the 2 numbers in the hexagons below it. Complete the puzzle. Take the number of the top hexagon and add 20 to it, what value do you get?

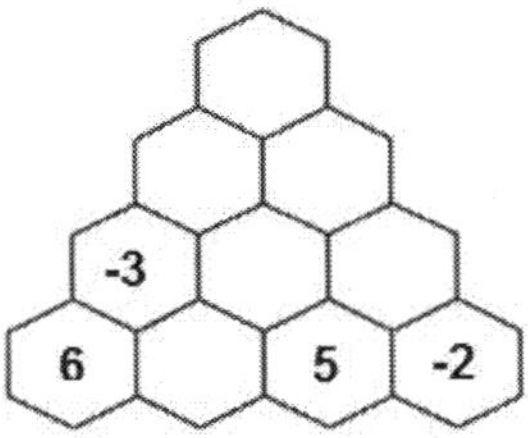
</td><td>12</td></tr>
<tr><td>6) For each of the numbers: 41, 53, and 82, the tens digit is greater in value than the ones digit. How many 2-digit numbers have this property, but exclude a 6 in either digit?</td><td>36</td></tr>
</table>

PERENNIAL MATH

TEST # 1

November 2015 Intermediate Level Solutions

We are only providing some suggested solutions below. There are always multiple methods to solve problems.

1) The missing number is 4. Both sides are equal to 26.

$$30 - \boxed{4} = 22 + \boxed{4}$$

4

2) Since Rachel has twice as much money as Kim and Kim has two dimes and two nickels (30 cents), then Rachel has to have 60 cents. Rachel only has quarters and nickels and has the same number of each coin, so she must have 2 quarters and 2 nickels. They trade all their money in for only nickels, so they have 90 cents altogether and that is 18 nickels.

18

3) Milo spent half of his weekly allowance for a movie ticket. His parents let him wash the car for $15 to earn more money. He ended up with $24. Work backwards.

$24 – $15 = $9

$9 x 2 = $18

His weekly allowance is $18.

18

PERENNIAL MATH

TEST # 1

November 2015 Intermediate Level Solutions

4) The sum of the closest palindrome before 2002 and the closest palindrome after 2002 is: 1991 + 2112 = 4103.

4103

5) If you complete the puzzle, -8 is in the top hexagon. -8 + 20 = 12

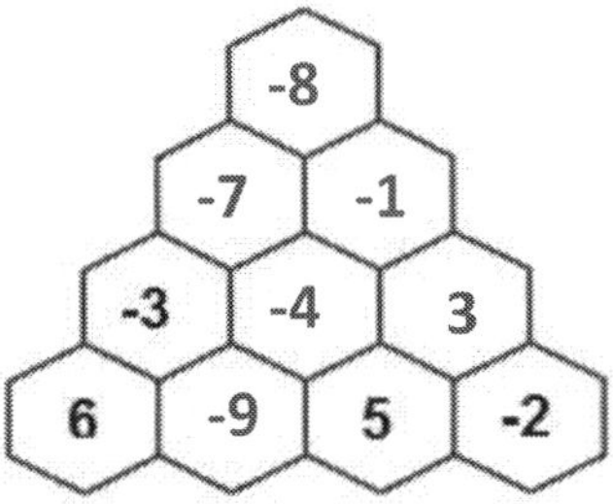

12

6) If we begin to list the numbers in groups: 10; 20,21; 30,31,32; 40,41,42,43; ... ; 90,91,92,93,94,95,96,97,98. We can see that the total number of 2-digit numbers, for which the ones digit is greater than the tens digit, will be 1 + 2 + ... + 9 = 45. Now subtract the nine 2-digit numbers that have a 6 in either digit. 45 – 9 = 36 numbers.

10								
20	21							
30	31	32						
40	41	42	43					
50	51	52	53	54				
60	61	62	63	64	65			
70	71	72	73	74	75	76		
80	81	82	83	84	85	86	87	
90	91	92	93	94	95	96	97	98

36

Name________________________________

PERENNIAL MATH

December 2015 Intermediate Level

TEST #2

1) 11 x 30 = 22 x □

2) Riley is half as old as Henry. The sum of their ages is 72. How old is Henry going to be next year?

3) The high school band is traveling. Five band members play the flute. There are three times as many members who play the trumpet than flutists. There are eight fewer trombone players than trumpeters, and eleven more drummers than trombone players. There are twice as many members that play the clarinet as members that play the flute. There are four fewer tuba players than there are trombone players, but three more members play the French horn than play the trombone. The band director, his assistant, and six parent volunteers are also going. How many seats are needed on the bus?

4) If 10★equals 12 ☾and 3☾equals 5♥, then 16 ♥ equals how many★ ?

5) How many triangles of any size are present in this drawing?

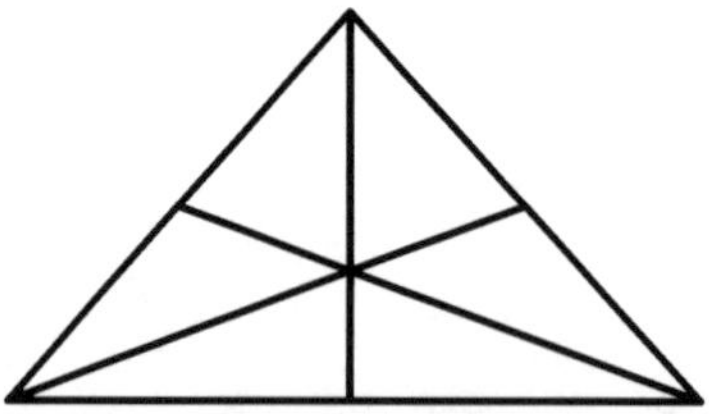

6) The sum of three consecutive odd numbers is 105. What is the smallest of the three numbers?

PERENNIAL MATH TEST #2

December 2015 Intermediate Level Answers

1) 11 x 30 = 22 x ☐	15
2) Riley is half as old as Henry. The sum of their ages is 72. How old is Henry going to be next year?	49
3) The high school band is traveling. Five band members play the flute. There are three times as many members who play the trumpet than flutists. There are eight fewer trombone players than trumpeters, and eleven more drummers than trombone players. There are twice as many members that play the clarinet as members that play the flute. There are four fewer tuba players than there are trombone players, but three more members play the French horn than play the trombone. The band director, his assistant, and six parent volunteers are also going. How many seats are needed on the bus?	76
4) If 10★ equals 12☾ and 3☾ equals 5♥, then 16♥ equals how many★ ?	8
5) How many triangles of any size are present in this drawing? 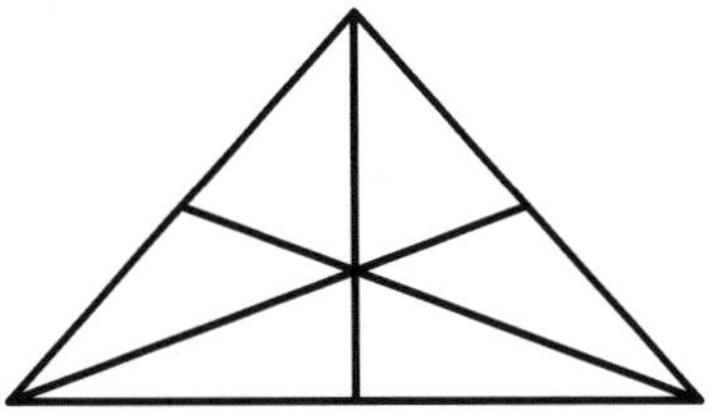	16
6) The sum of three consecutive odd numbers is 105. What is the smallest of the three numbers?	33

PERENNIAL MATH

TEST # 2

December 2015 Intermediate Level Solutions

We are only providing some suggested solutions below. There are always multiple methods to solve problems.

1) 11 x 30 = 22 x ☐

11 x 30 = 330
330 ÷ 22 = 15

15

2) Riley is half as old as Henry. The sum of their ages is 72. How old is Henry going to be next year?

Let R stand for Riley's age.

R + 2R = 72
3R = 72
R = 24.

If Riley is 24, then Henry is 48. Next year Henry will be 49.

49

3) Break out each instrument and add together:

Five members play the flute	5 flutes
Three times as many trumpet players than flutists	15 trumpets
Eight fewer trombone players than trumpeters	7 trombones
Eleven more drummers than trombone	18 drummers
Twice as many clarinets as flutes	10 clarinets
Four fewer tuba players than there are trombone	3 tubas
Three more French horn than trombone	10 French Horns
	1 director
	1 assistant
	6 parents
	76 seats are needed

76

PERENNIAL MATH

TEST #2

December 2015 Intermediate Level Solutions

4) If 10★equals 12 ☾ and 3☾ equals 5♥, then 12 ☾equals 20 ♥ which also makes 10 ★ equal 20 ♥. So there is a 1★ to 2 ♥ ratio. Therefore, 16 ♥ would be equal to 8★ .

8

5)

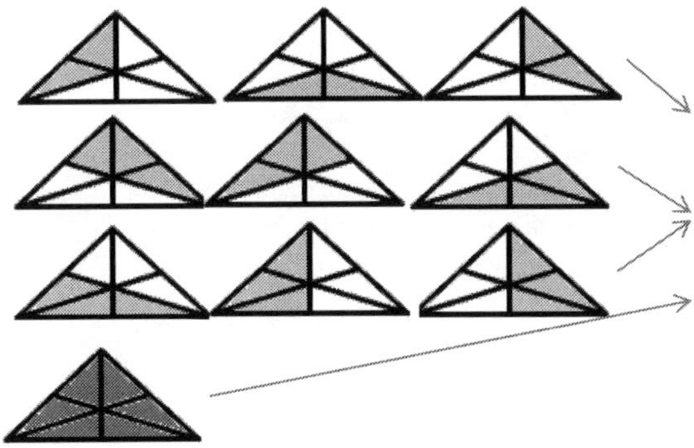

Type of triangle	How many each type
Made from 1 shape	6
Made from 2 shapes	3
Made from 3 shapes	6
Made from 6 shapes	1
Total	16

16

6) The sum of three consecutive odd numbers is 105.
Divide 105 by 3 to find the average or middle number:

105 / 3 = 35

The three consecutive odd numbers are 33, 35, and 37.
Therefore, the smallest number is 33.

33

Name________________________________

PERENNIAL MATH

TEST #3

January 2016 Intermediate Level

1) Sam is thinking of a number. If you subtract 4 from his number and then multiply by 7, the answer is 56. What number is Sam thinking of?

2) If Darla makes 3 bells the first day of the month, 5 bells the second day, 8 bells the third day and then starts to repeat this pattern again, how many bells will she have made at the end of January?

3) Bayside Kayak Shop charges a 15 dollar fee plus an additional 8 dollars an hour for renting a kayak. Nick paid 87 dollars to rent a kayak. How many hours did he rent the kayak for?

4) The Rangers are one of 7 teams in their division. They play each team in their division 15 games each. However, they also play each of the 8 teams in the western division an equal number of games. If they play a total of 146 games for the season, how many games do they play against each of the 8 western division teams?

5) This 3 x 3 grid is made up of a border of eight squares surrounding one square. A 3 x 4 grid would be made up of a border of ten squares surrounding two squares. How many squares would be in the border of a 3 x 37 grid?

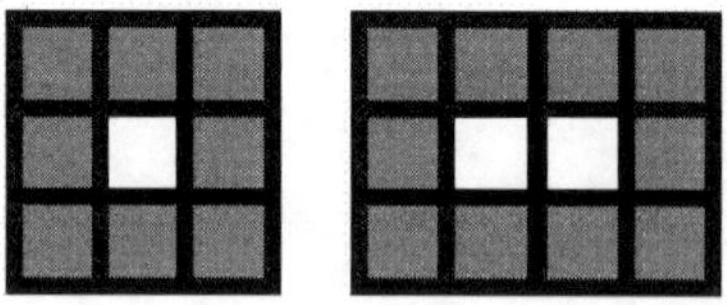

6) The ratio of boys to girls at the movies is 5:3. There are 36 girls. How many more boys are there than girls?

PERENNIAL MATH

TEST #3

January 2016 Intermediate Level Answers

1) Sam is thinking of a number. If you subtract 4 from his number and then multiply by 7, the answer is 56. What number is Sam thinking of?	12
2) If Darla makes 3 bells the first day of the month, 5 bells the second day, 8 bells the third day and then starts to repeat this pattern again, how many bells will she have made at the end of January?	163
3) Bayside Kayak Shop charges a 15 dollar fee plus an additional 8 dollars an hour for renting a kayak. Nick paid 87 dollars to rent a kayak. How many hours did he rent the kayak for?	9
4) The Rangers are one of 7 teams in their division. They play each team in their division 15 games. However, they also play each of the 8 teams in the western division an equal number of games. If they play a total of 146 games for the season, how many games do they play against each of the 8 western division teams?	7
5) This 3 x 3 grid is made up of a border of eight squares surrounding one square. A 3 x 4 grid would be made up of a border of ten squares surrounding two squares. How many squares would be in the border of a 3 x 37 grid? 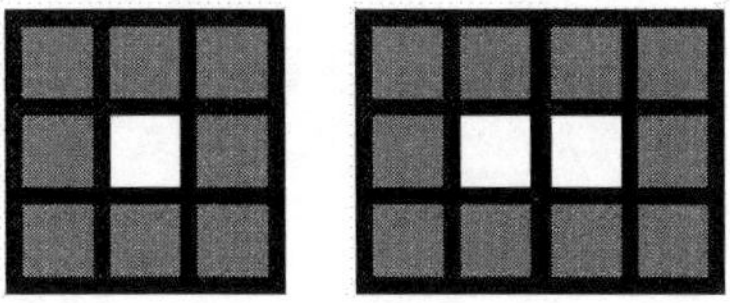	76
6) The ratio of boys to girls at the movies is 5:3. There are 36 girls. How many more boys are there than girls?	24

PERENNIAL MATH TEST #3

January 2016 Intermediate Level Solutions

We are only providing some suggested solutions below. There are always multiple methods to solve problems.

1) Sam is thinking of a number. If you subtract 4 from his number and then multiply by 7, the answer is 56.

Work backwards. 56 ÷7 = 8.
Now add 4. 8 +4 = 12. His number was 12.
You can work it forward to check. (12 – 4) x 7 = 56

12

2) If Darla makes 3 bells the first day of the month, 5 bells the second day, 8 bells the third day and then starts to repeat this pattern again, how many bells will she have made at the end of January?

You must first know that there are 31 days in January. So there will be 10 sets of the pattern 3 + 5 + 8 = 16, and then 1 extra day which she will make 3 bells. (10 x 16) + 3 = 163

163

3) Bayside Kayak Shop charges a 15 dollar fee plus an additional 8 dollars an hour for renting a kayak. Nick paid 87 dollars to rent a kayak. How many hours did he rent the kayak for?

First subtract the original 15 dollar fee. $87 - $15 = $72 Now divide that amount by $8 per hour. $72/8 = 9. Nick rented the kayak for 9 hours.

9

PERENNIAL MATH

TEST #3

January 2016 Intermediate Level Solutions

4) The Rangers are one of 7 teams in their division. They play each team in their division 15 games. However, they also play each of the 8 teams in the western division an equal number of games. If they play a total of 146 games for the season, how many games do they play against each of the 8 western division teams?
They play the other 6 teams in their division 15 games. 6 x 15 = 90. Subtract 90 from 146 and they have 56 games. Since there are 8 teams in the western division, 56 games divided by 8 teams equals 7 games against each of the western teams.

7

5) This 3 x 3 grid is made up of a border of eight squares surrounding one square. A 3 x 4 grid would be made up of a border of ten squares surrounding two squares. How many squares would be in the border of a 3 x 37 grid?

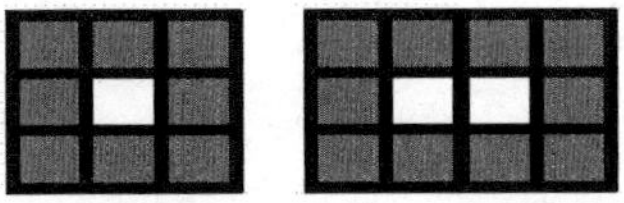

Let n stand for the number that we are multiplying 3 by to make the grid.
The pattern for this is $3 \times n - (n - 2)$
$3 \times 37 - (37-2) = 111 - 35 = 76$.
There are 76 squares in the border of a 3 x 37 grid.

76

6) The ratio of boys to girls at the movies is 5:3. There are 36 girls. How many more boys are there than girls? There are 3 portions of girls for a total of 36 girls. 1 portion would equal 12 girls. Since there are 5 portions of boys, then 5 x 12 = 60 boys. 60 boys – 36 girls = 24 boys. There are 24 more boys at the movies than girls.

24

Name________________________________

PERENNIAL MATH

TEST #4

February 2016 Intermediate Level

1) Jim made 4 stacks of coins. The first stack had 6 coins. The second stack had 2 times as many coins as the first. The third stack was 3 coins less than the second stack, and the last stack was 10 coins more than the first stack. How many more coins did the third and fourth stack have compared to the first and second stack ?	
2) Eli has $148 in the bank. He deposits $8 at the end of each week. Peyton has $260 in the bank. He withdraws $6 at the end of each week. How many weeks will it take for them to have the same amount in the bank?	
3) Mr. Tillman put $6,000 in an investment account. He earned 10% interest each year. If he left that money in the account for two years earning him cumulative interest, how much money will he have in the account at the end of those two years?	
4) Michael lifts weights every 6 days, runs every 4 days and swims every 16 days. If he did all three activities today, in how many days will he do all three activities again on the same day?	
5) If the pattern here were to be continued five more places, what would be the total sum of all the tiny black squares in the series? 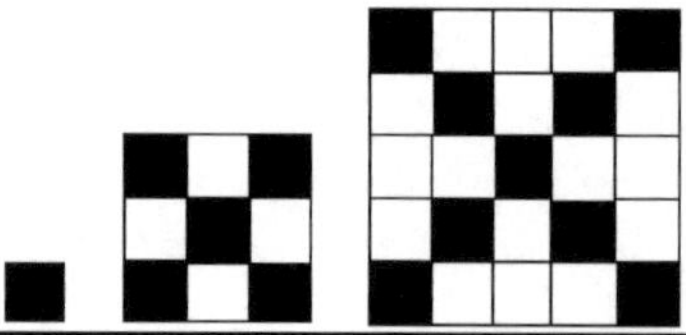	
6) How many three-digit numbers between 100 and 999 are multiples of 27 and do not contain the digit 8?	

PERENNIAL MATH

TEST #4

February 2016 Intermediate Level Answers

1) Jim made 4 stacks of coins. The first stack had 6 coins. The second stack had 2 times as many coins as the first. The third stack was 3 coins less than the second stack, and the last stack was 10 coins more than the first stack. How many more coins did the third and fourth stack have compared to the first and second stack ?	7
2) Eli has $148 in the bank. He deposits $8 at the end of each week. Peyton has $260 in the bank. He withdraws $6 at the end of each week. How many weeks will it take for them to have the same amount in the bank?	8
3) Mr. Tillman had $6,000 in his retirement account. He earned 10% interest each year. If he left that money in the account for two years, how much money will he have in the account at the end of those two years? Remember to apply cumulative interest.	7260
4) Michael lifts weights every 6 days, runs every 4 days and swims every 16 days. If he did all three activities today, in how many days will he do all three activities again on the same day?	48
5) If the pattern here were to be continued five more places, what would be the total sum of all the tiny black squares in the series? 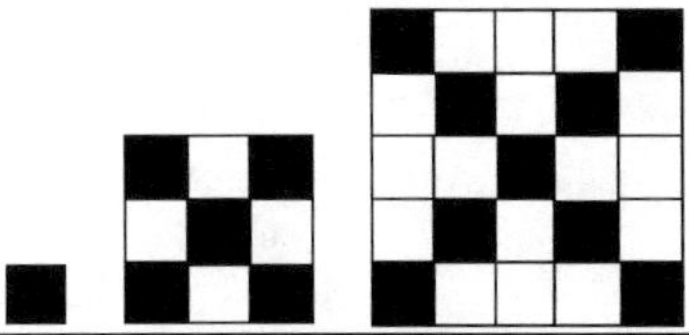	120
6) How many three-digit numbers between 100 and 999 are multiples of 27 and do not contain the digit 8?	22

PERENNIAL MATH

TEST # 4

February 2016 Intermediate Level Solutions

We are only providing some suggested solutions below. There are always multiple methods to solve problems.

1) To find out how many more coins are in the last two stacks compared to the first two stacks, list the stacks and add the first two and subtract the them from the sum of the last two. Stack 1 = 6. Stack 2 is twice that many which = 12. The sum of the first two stacks is 6 + 12 = 18. Stack 3 is 3 less than stack 2 or 12 - 3 = 9. Stack 4 is 10 more than stack 1 or 6 + 10 = 16. The sum of stack 3 and 4 is 9 + 16 = 25. Now subtract the two sums 25 – 18 = 7.

7

2) Make a chart to track the accounts of each boy. Then you will be able to see when they have the same amount of money and how long it took them to get the same amount. It takes 8 weeks of transactions before Eli and Peyton both have the same amount.

	Start	Week 1	Week 2	Week 3	Week 4	Week 5	Week 6	Week 7	Week 8
Eli (+8)	$148	$156	$164	$172	$180	$188	$196	$204	$212
Peyton (-6)	$260	$254	$248	$242	$236	$230	$224	$218	$212

8

3) Mr. Tillman started with $6,000. At the end of 1 year he gains 10% interest which is 6000 x 10% = $600. His account balance at the end of the first year is $6,600 in his retirement account. The second year he would earn $6600 x 10% = $660. His account balance at the end of the second year would be $6600 + $660 = $7260.

	Starting Balance	10 % Interest	Ending Balance
Year 1	$6,000	$600	$6,600
Year 2	$6,600	$660	$7,260

7,260

PERENNIAL MATH

TEST #4

February 2016 Intermediate Level Solutions

4) Find the lowest common multiple of 4, 6, and 16:

Multiples of 4: 4, 8, 12, 16, 20, 24, 28, 32, 36, 40, 44, **48**, 52.

Multiples of 6: 6, 12, 18, 24, 30, 36, 42, **48**, 54.

Multiples of 16: 16, 32, **48**

In 48 days he will do all three activities on the same day.

48

5) In the pattern each new item has four more black squares than the previous item in the series. If the pattern here were to be continued five more places, the total sum of all the tiny black squares in the series would be:

1 + 5 + 9 + 13 + 17 + 21 + 25 + 29 = 120.

Smart addition can be applied (1 + 29) + (5 + 25) + (9 + 21) + (13 + 17) =

30 + 30 + 30 + 30 = 120

120

6)Make a list of the three-digit multiples of 27. There are 33. Since the problem stated between 100 and 999, do not count 100 or 999. Also mark out any of the multiples that contain the digit 8. There are 11 multiples with the digit 8.

33 – 11 = 22 multiples that do not include the digit 8.

Multiples include:

~~108~~,135,162,~~189~~,216,243,270,297,324,351,~~378~~,405,432,459,~~486~~,513,540,567,594,621, ~~648~~,675,702,729,756,~~783~~,~~810~~,~~837~~,~~864~~,~~891~~,~~918~~,945,972. Do not count 999.

22

Name________________________________

PERENNIAL MATH

TEST #1

November 2016 Intermediate Level

1) The product of three different whole numbers is 80. The sum of the three numbers is 15. What is the difference between the largest and smallest number?

2) What is the sum of the series 1+2+3+4+5+6+7+8+9 · · ·+18+19+20+21

3) 3 + 7 and 5 + 5 are two ways to express 10 as the sum of two prime numbers. In how many ways can 40 be expressed as the sum of two prime numbers?

4) Use the even numbers 2-12 to fill in the circles so that the three sums along each side of triangle equals 20. What is the sum of the three corner circles?

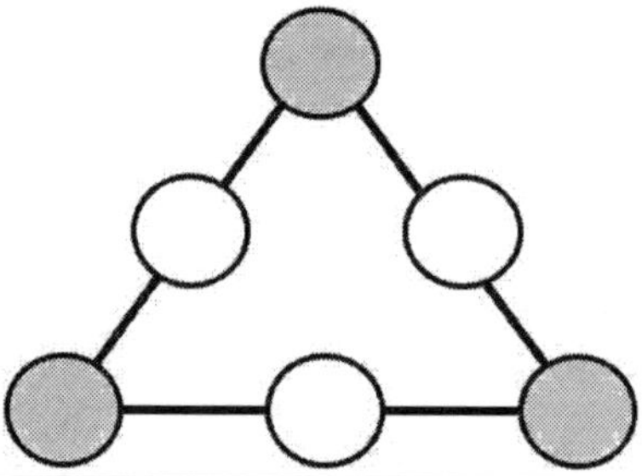

5) There are 128 tennis players participating in a tennis tournament. Each player has only one match to play per day. Every player plays on the first day, but only winners continue to play the next day. If the tournament started on August 29th, what would be the date of the final match where only 2 players are left to play?

6) The sides of a square are 12 inches long. If you increase each side by 50% by how many square inches will the area of the square grow?

PERENNIAL MATH

TEST # 1

November 2016 Intermediate Level Answers

1) The product of three different whole numbers is 80. The sum of the three numbers is 15. What is the difference between the largest and smallest number?	6
2) What is the sum of the series 1+2+3+4+5+6+7+8+9 · · ·+18+19+20+21	231
3) 3 + 7 and 5 + 5 are two ways to express 10 as the sum of two prime numbers. In how many ways can 40 be expressed as the sum of two prime numbers?	3
4) Use the even numbers 2-12 to fill in the circles so that the three sums along each side of triangle equals 20. What is the sum of the three corner circles? 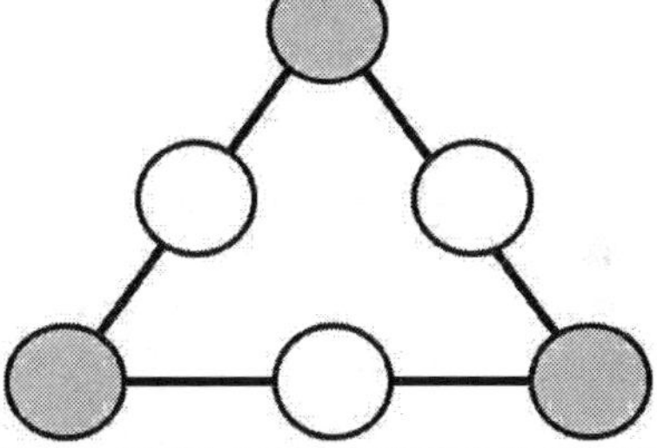	18
5) There are 128 tennis players participating in a tennis tournament. Each player has only one match to play per day. Every player plays on the first day, but only winners continue to play the next day. If the tournament started on August 29th, what would be the date of the final match where only 2 players are left to play?	Sept 4
6) The sides of a square are 12 inches long. If you increase each side by 50% by how many square inches will the area of the square grow?	180

PERENNIAL MATH

TEST # 1

November 2016 Intermediate Level Solutions

We are only providing some suggested solutions below. There are always multiple methods to solve problems.

1) The product of three different whole numbers is 80. The sum of the three numbers is 15. To find the difference between the largest and smallest number first list numbers that can be multiplied to yield 80. Since 80 ends in zero, it is a multiple of 0 or 5. Since it can't be a multiple of 0 in this case, 5 has to be one of the numbers. Therefore, 5, 2, and 8 are the three numbers because they add up to 15 and their product is 80. The difference between the largest and smallest number is 8-2 = 6.

6

2) To find the sum of the series 1+2+3+4+5+6+7+8+9 · · · +18+19+20+21, do some smart addition. If you pair up numbers (1,20), (2, 19), (3, 18), (4, 17), (5, 16), (6, 15), (7, 14), (8, 13), (9, 12), (10,11). That is ten pairs of 21 plus we have the number 21 to add. So (10 x 21) + 21 = 210 + 21= 231

231

3) In how many ways can 40 be expressed as the sum of two prime numbers? List the prime numbers and then do an organized list to make sure you do not omit or repeat any combinations. The prime numbers to 40 include 2, 3, 5, 7, 11, 13, 17, 19, 23, 29, 31, 37. The sums that work are (3 + 37), (11 + 29) and (23 +17). There are 3 ways to make 40.

3

PERENNIAL MATH TEST #1

November 2016 Intermediate Level Solutions

4) First make a list of the combinations of even numbers 2-12 that add up to 20. Select numbers from the combinations that appear twice and place those in the corners. The numbers 2, 6, and 10 appear in corners and their sum is 18.

Combinations with sum of 20		
2	6	12
2	8	10
4	6	10

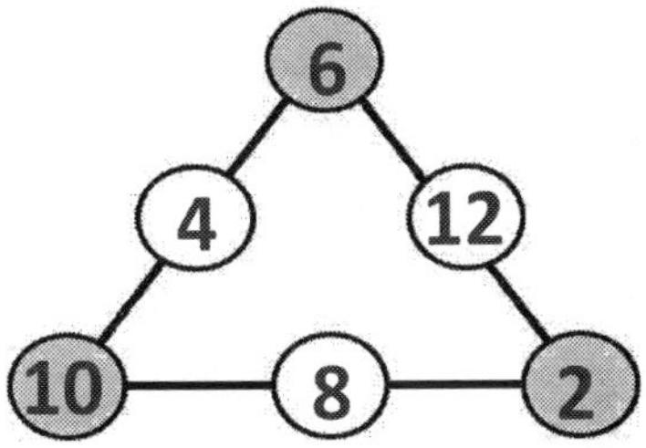

18

5) There are 128 tennis players with one match per day, you can make a chart. The first matches took place on August 29th and there would be 128 ÷ 2 = 64 players left. On August 30th, those players would play and 64 ÷ 2 = 32 players left. On August 31st, those players would play and 32 ÷ 2 = 16 players left. On Sept 1st, those player would play and 16 ÷ 2 = 8 players left. On Sept 2nd, 8 ÷ 2 = 4 players left. On Sept 3rd, 4 ÷ 2 = 2 players left. The final two players will play for the championship on Sept 4th.

Date	Aug 29th	Aug 30th	Aug 31st	Sept 1st	Sept 2nd	Sept 3rd	Sept 4th
Players Remaining	128	64	32	16	8	4	2

Sept 4th

6) The sides of a square are 12 inches long and you increase each side by 50%. To find how many square inches will the area will grow by first solve for the original area which is 12^2 = 144 square inches. If both sides are increased by 50% the new area would increase by 6 on each side and be 18^2 or 324 square inches. Now subtract the original area 324 – 144 = 180 square inches.

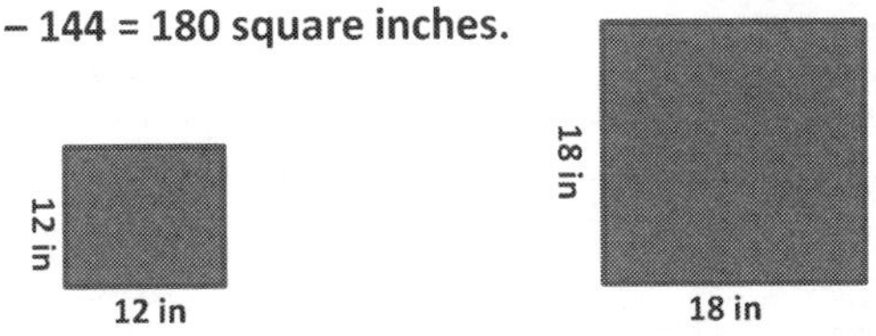

180

Name________________________________

PERENNIAL MATH

TEST #2

December 2016 Intermediate Level

1) The product of two whole numbers is 72. The average of the two numbers is 9. What is the largest of the two numbers?	
2) The sum of five consecutive whole numbers is 2015. What is the smallest number among the five numbers?	
3) When a positive whole number is divided by 7, the remainder is 2. When the same number is divided by 5, the remainder is also 2. When the positive number is divided by 8, there is no remainder. What is the least possible number?	
4) This large cube is made from smaller 1 x 1 inch cubes. Cubes have been removed so that you can see straight through every side. How many 1 x 1 cubes are left in this shape? 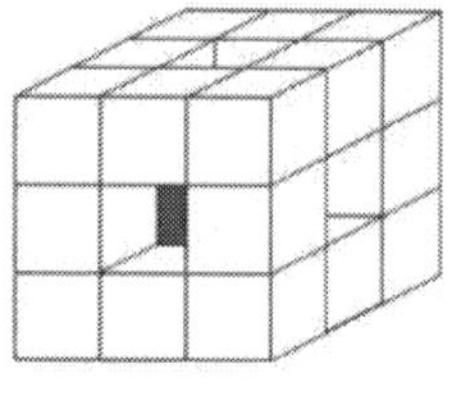	
5) By exchanging the positions of two digits in the number 965,142 the new number is 19,980 smaller than the original number. What is the sum of the two digits exchanged?	
6) To run 1500 meters in John's rectangular training facility, John could run the length 50 times or he could run the perimeter of the facility 20 times. In square meters, what is the area of the facility?	

PERENNIAL MATH

TEST #2

December 2016 Intermediate Level Answers

1) The product of two whole numbers is 72. The average of the two numbers is 9. What is the largest of the two numbers?	12
2) The sum of five consecutive whole numbers is 2015. What is the smallest number among the five numbers?	401
3) When a positive whole number is divided by 7, the remainder is 2. When the same number is divided by 5, the remainder is also 2. When the positive number is divided by 8, there is no remainder. What is the least possible number?	72
4) This large cube is made from smaller 1 x 1 inch cubes. Cubes have been removed so that you can see straight through every side. How many 1 x 1 cubes are left in this shape? 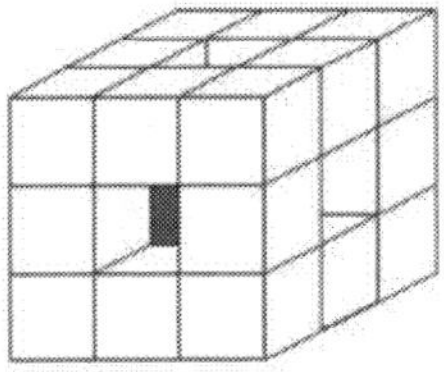	20
5) By exchanging the positions of two digits in the number 965,142 the new number is 19,980 smaller than the original number. What is the sum of the two digits exchanged?	10
6) To run 1500 meters in John's rectangular training facility, John could run the length 50 times or he could run the perimeter of the facility 20 times. In square meters, what is the area of the facility?	225

PERENNIAL MATH

TEST #2

December 2016 Intermediate Level Solutions

We are only providing some suggested solutions below. There are always multiple methods to solve problems.

1) Since the product of two whole numbers is 72 we can list out the factors of 72 (1, 2, 3, 4, 6, 8, 9, 12, 18, 24). Then we can select two numbers whose sum is 18 because their average is 9 and the formula for average is the sum of all items divided by the number of items. The two numbers are 12 and 6 so the largest number is 12.

12

2) The sum of five consecutive whole numbers is 2015. To find the smallest number among the five numbers you can first find the average of the numbers which is the middle number by dividing by five. 2015 ÷ 5 = 403. So the five numbers are 401, 402, 403, 404, and 405. The smallest possible number is 401.

401

3) When a positive whole number is divided by 7, the remainder is 2. When the same number is divided by 5, the remainder is also 2. When the positive number is divided by 8, there is no remainder. To find the least possible number make a table to see when the numbers match up. 72 is the least possible number that appears in all 3 rows.

Counts by 5 with 2 left over	7	12	17	22	27	32	37	42	47	52	57	62	67	72
Counts by 7 with 2 left over	9	16	23	30	37	44	51	58	65	72				
Counts by 8 with 0 left over	8	16	24	32	40	48	56	64	72					

72

PERENNIAL MATH

TEST #2

December 2016 Intermediate Level Solutions

4) This large cube is made from smaller 1 x 1 inch cubes. Cubes have been removed so that you can see straight through every side. There are 3 x 9 = 27 cubes that make this shape when solid and one cube would be removed from each of the sides and one from the center to see straight through every side. That would leave 27 – 7 = 20 cubes remaining in the shape.

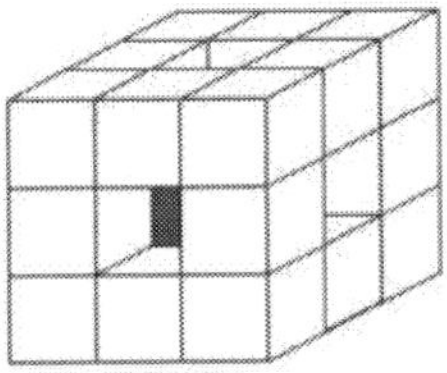

20

5) By exchanging the positions of two digits in the number 965,142 the new number is 19,980 smaller than the original number. To find the sum of the two digits exchanged subtract the difference and you'll be able to notice a new number formed. The 6 and 4 are exchanged and their sum is 10.

$$\begin{array}{r} 965{,}142 \\ -19{,}980 \\ \hline 945{,}162 \end{array}$$

10

6) To run 1500 meters in John's rectangular training facility, John could run the length 50 times or he could run the perimeter of the facility 20 times. In square meters, what is the area of the facility? First you can solve for the length of facility by dividing the 1500 meters by the 50 times, or 1500/50 = 30 meters. Next you can solve for the perimeter by dividing 1500 meters by the 20 times, 1500/20 = 75 meters. Now solve for the width by doubling the length (30 x 2 = 60) and subtracting that from the perimeter (75 – 60 = 15 meters) and then dividing by 2 (15/2 = 7.5). Therefore, the length = 30 meters and the width = 7.5 meters. The area of the facility is equal to the length times the width, 30 x 7.5 = 225 m^2

225

Name________________________________

PERENNIAL MATH

January 2017 Intermediate Level — TEST #3

1) 297 x 35 + 297 x 65

2) The average of Ben's three tests is 89. Each test is worth 100 points. What score does he have to make on his 4th test to get an average of 90?

3) A painter mixes 4 gallons of red paint with 1 gallon of blue paint to make 5 gallons of her signature purple paint. Each gallon of red paint costs \$2 and each gallon of blue paint costs \$3. She only has \$670. How much money does the painter need to borrow to make 400 gallons of purple paint?

4) Find the difference between the largest and smallest of these three fractions a, b, and c. What is the greatest common factor of the answer before the fraction is reduced?

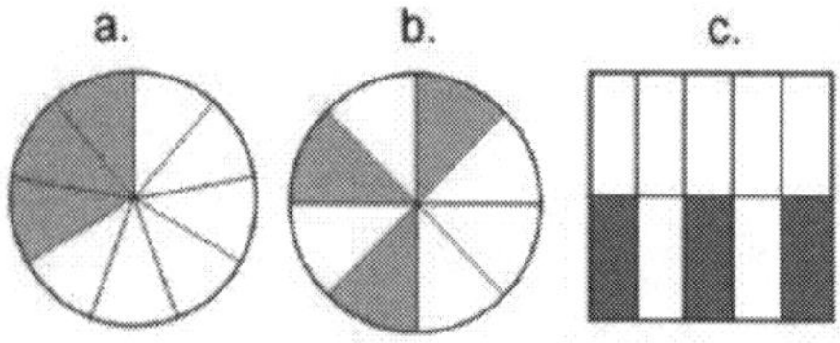

5) If $\frac{N}{30}$ is a simplified fraction (that is, reduced to lowest terms) and is between 0 and 1, how many values are possible for N?

6) What is the digit in the ones place of the following? (7 × 7 × 7 × 7 × 7) × (7 × 7 × 7 × 7 × 7) × (7 × 7 × 7 × 7 × 7) × (7 × 7 × 7 × 7 × 7)

PERENNIAL MATH TEST #3

January 2017 Intermediate Level Answers

1) 297 x 35 + 297 x 65	29,700
2) The average of Ben's three tests is 89. Each test is worth 100 points. What score does he have to make on his 4th test to get an average of 90?	93
3) A painter mixes 4 gallons of red paint with 1 gallon of blue paint to make 5 gallons of her signature purple paint. Each gallon of red paint costs $2 and each gallon of blue paint costs $3. She only has $670. How much money does the painter need to borrow to make 400 gallons of purple paint?	210
4) Find the difference between the largest and smallest of these three fractions a, b, and c. What is the greatest common factor of the answer before the fraction is reduced? 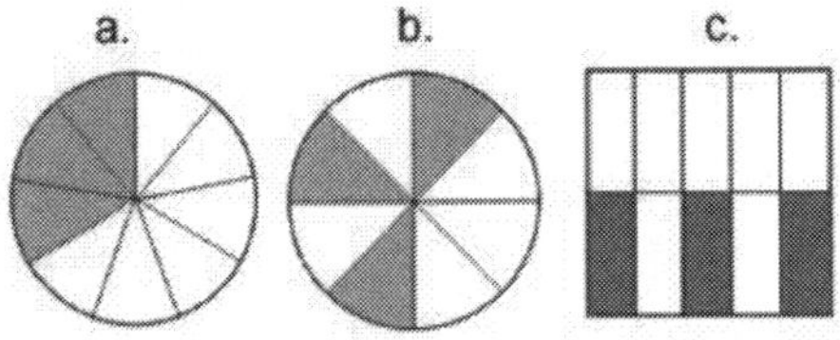	9, 1, or 6
5) If $\frac{N}{30}$ is a simplified fraction (that is, reduced to lowest terms) and is between 0 and 1, how many values are possible for N?	8
6) What is the digit in the ones place of the following? (7 × 7 × 7 × 7 × 7) × (7 × 7 × 7 × 7 × 7) × (7 × 7 × 7 × 7 × 7) × (7 × 7 × 7 × 7 × 7)	1

PERENNIAL MATH

TEST #3

January 2017 Intermediate Level Solutions

We are only providing some suggested solutions below. There are always multiple methods to solve problems.

1) Use smart multiplication, rearrange the numbers:

297 x (35 + 65) = 297 x 100 = 29700

29,700

2) Average = $\dfrac{\textbf{Sum of all Tests}}{\textbf{Number of Tests}}$

Ben's average score on three test is 89. Multiply 89 x 3 so the total sum of his tests is 267. To make an average of 90 he will have to have a total sum of 4 tests to be equal to (90 x 4) or 360. Subtract his current sum and you will have what he needs to make. 360 – 267 = 93. He will need to score 93 points.

93

3) A painter mixes 4 gallons of red paint with 1 gallon of blue paint to make 5 gallons of her signature purple paint. Each gallon of red paint costs $2 and each gallon of blue paint costs $3. She only has $670. To find out how much money the painter needs to borrow to make 400 gallons of purple paint convert the ratios of paint and cost. For each 5 gallons there 4:1 ratio of red to blue paint. The cost is $8 + $3 or $11 per 5 gallons. Divide 400 by 5 and you will find out how batches will be made. 80 batches of purple can be made at $11 per batch 80 x $11 = $880. Subtract to see how much money she will need to borrow. $880 - $670 = $210, She will need to borrow $210.

$8 + $3 = $11 per 5 gallons

210

PERENNIAL MATH

TEST #3

January 2017 Intermediate Level Solutions

4) To find the difference between the largest and smallest of these three fractions a, b, and c, convert them to like terms by finding the least common multiple. The least common multiple of 8, 9, and 10 is 360. The smallest fraction is 108/360. The largest is 135/360. The difference of the largest and smallest is 135/360 – 108/360 = 27/360. The greatest common factor in the numerator and denominator of the answer is 9.

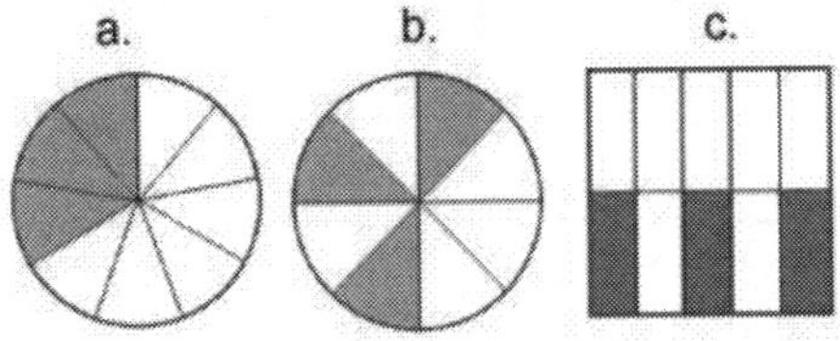

a) 3/9 = 120/360
b) 3/8 = 135/360
c) 3/10 = 108/360

- Some students automatically recognize that 3/8 is the largest and 3/10 is the smallest fraction. Their common denominator would be 40. Therefore, 3/8 – 3/10 = 15/40 – 12/40 = 3/40. The greatest common factor in the numerator and denominator would be 1.
- If students interpreted fraction a as 1/3 then the common denominator would have been 240. then 90/240 – 72/240 = 18/240. The greatest common factor in the numerator and denominator would be 6.

9, 1, 6

5) If $\frac{N}{30}$ is a simplified fraction between 0 and 1, the values are possible for N include: 1/30, 7/30, 11/30, 13/30, 17/30, 19/30, 23/30, 29/30. There are 8 possible values for N.

8

6) To find the digit in the ones place of the following: (7 × 7 × 7 × 7 × 7) × (7 × 7 × 7 × 7 × 7) × (7 × 7 × 7 × 7 × 7) × (7 × 7 × 7 × 7 × 7), Look for a pattern in the multiplication. The ones place has a pattern of 7, 9, 3, 1. Since this problem is 7^{20} the ones digit would be 1.

	7	49	343	2401	16807	117649	823543	5764801
Ones digit	7	9	3	1	7	9	3	1

1

Name________________________________

PERENNIAL MATH

TEST #4

February 2017 Intermediate Level

1) What is the sum of the different prime factors of 60?	
2) Trixie is 80% heavier than Rover. Scooby is 80% lighter than Rover. If Rover weighs 40 pounds, how much do all three dogs weigh altogether?	
3) Milo divided his coins into two piles. The two piles had an equal number of coins. After Milo gave his sister two-thirds of one pile, he had 24 coins left. How many coins did Milo have at the start?	
4) Find the sum of the two numbers that should go in the boxes to make these two equations true. $5 = \frac{35}{\square} = \frac{\square}{15}$	
5) There are three clocks that chime every 4 minutes, 7 minutes, and 13 minutes, respectively. If all three clocks chime together at 3:15 p.m., when will they chime all together next time?	
6) The 5th grade class completed a final math test. The average score of the 18 students in Mrs. Black's class was 75, and the average score of the 16 students in Mrs. Smith's class was 92. What is the average test score of the entire 5th grade class?	

PERENNIAL MATH

TEST #4

February 2017 Intermediate Level Answers

1) What is the sum of the different prime factors of 60?	10
2) Trixie is 80% heavier than Rover. Scooby is 80% lighter than Rover. If Rover weighs 40 pounds, how much do all three dogs weigh altogether?	120
3) Milo divided his coins into two piles. The two piles had an equal number of coins. After Milo gave his sister two-thirds of one pile, he had 24 coins left. How many coins did Milo have at the start?	36
4) Find the sum of the two numbers that should go in the boxes to make these two equations true. $5 = \frac{35}{\square} = \frac{\square}{15}$	82
5) There are three clocks that chime every 4 minutes, 7 minutes, and 13 minutes, respectively. If all three clocks chime together at 3:15 p.m., when will they chime all together next time?	9:19 p.m.
6) The 5th grade class completed a final math test. The average score of the 18 students in Mrs. Black's class was 75, and the average score of the 16 students in Mrs. Smith's class was 92. What is the average test score of the entire 5th grade class?	83

PERENNIAL MATH

TEST #4

February 2017 Intermediate Level Solutions

We are only providing some suggested solutions below. There are always multiple methods to solve problems.

1) The factors of 60 are 1, 60, 2, 30, 3, 20, 4, 15, 5, 12, 6, 10. The prime factors are 2, 3, and 5. Their sum is 10

10

2) Rover weighs 40 pounds and Trixie is 80% heavier so Trixie weighs 40 ÷ 80% = 32 more or 72 pounds. Scooby is 80% lighter than Rover. So Scooby weighs 8 pounds. All three dogs weigh 40 + 72 + 8 = 120 pounds altogether.

120

3) Work backwards. Milo's two piles had an equal number of coins before he gave his sister two-thirds of one pile. He had 24 coins left. If we convert his piles to fractions he started with 6/3. So if he gave 2/3 away he would have 4/3 left. 24 = 4/3 That would make each (1/3) have 6 coins. He had 36 coins to start with.

Now you can test this working forwards: 36 ÷ 2 piles = 18 coins per pile. Milo gave 2/3 of one pile: 18 x (2/3) = 12. So there would be 18 - 12 = 6 coins left in the one pile. 6 + 18 in the other pile = 24 left. The original number of coins to start with is 36.

36

PERENNIAL MATH

TEST #4

February 2017 Intermediate Level Solutions

4) Find the sum of the two numbers that should go in the boxes to make these two equations true. The numbers that fit in the boxes are 7 and 75 because 35/7 and 75/15 both equal 5. The sum of the numbers in the boxes is 7 + 75 = 82.

$$5 = \frac{35}{\square} = \frac{\square}{15}$$

82

5) Since three clocks chime every 4 minutes, 7 minutes, and 13 minutes, respectively and all three clocks chime together at 3:15 p.m., the next time they will chime simultaneously is 364 minutes later because the least common multiple of 4, 7, and 13 is 364. Convert 364 to hours and minutes and add to 3:15 p.m. 364 minutes = 6 hours and 4 min. The time will be 9:19 p.m.

9:19 p.m.

6) The average score of the 18 students was 75, and the average score of the 16 students was 92. The average test score of the entire 5th grade class can be found by finding the total number of points scored by all students and dividing it by the total number of students. The total score from all the students:

(18 x 75) + (16 x 92) = 1350 + 1472 = 2822.

Divide the total score by the total number of students: 2822 ÷ 34 = 83

83

Made in the USA
Monee, IL
05 November 2026

67811021R00031